Match the Views

Geoff Giles

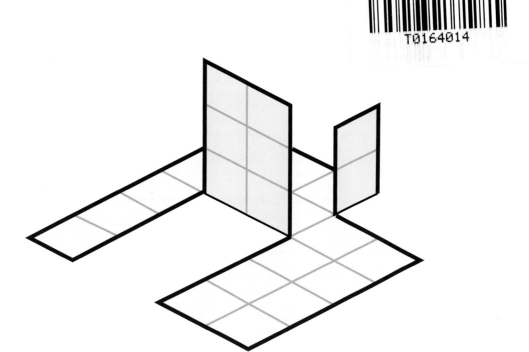

Tarquin Publications

Stimulating the Visual Imagination

All the puzzles in this book are based on imagining what a square of paper coloured yellow on one side and blue on the other is like when it is cut and folded. It offers plenty of opportunities for testing and improving your skill at visualising three-dimensional objects.

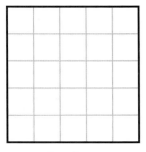

A 25 Square

The idea is to take a square five units by five, coloured yellow on one side and blue on the other.

Logically enough, It is is called a **25 Square** and in all the diagrams it is turned through 45° and squashed to be shown as an **isometric View**.

Isometric View

Using this standard isometric view of the 25 square, the cut and fold diagrams are drawn showing either the square before cutting, the **Plan View**, or after cutting and folding, the **Folded View**.

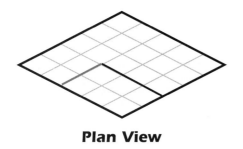

Plan View

On the **Plan View** the black lines show where the square is to be cut and red lines show where it is folded up at right angles to form a valley fold. The sets of puzzles L and M also include hill folds and such fold lines are coloured green.

The **Folded View** shows the result of cutting and folding. Remember that the underside of the sheet is coloured blue.

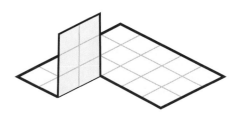

Folded View

A Sample Puzzle

All of the puzzles take the same form. On the left page of the spread are nine Plan Views and on the right side nine Folded Views. The puzzle and the challenge is to match each Plan View to its corresponding Folded View, simply by thinking and imagining the three-dimensional situation.

Plan View

Folded View

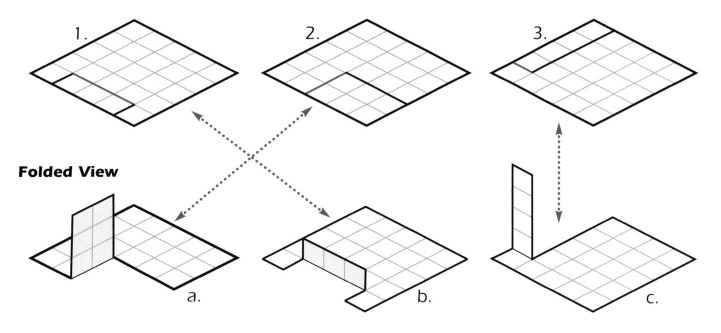

The answers can be entered into tables like these.

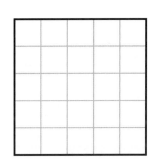

A	1	2	3
	b	a	c

This way round when you are going from Plan View to Folded View

A	a	b	c

This way round when you are going from to Folded View to Plan View

If you wish, you can also cut out and fold a real square of paper and check the answer directly. See pages 30 - 32 for some useful photocopy masters.

Puzzle A
Plan views

In your imagination cut and fold these squares and match them to the views opposite.

1.

2.

3.

4.

5.

6.

7.

8.

9.

A	1	2	3	4	5	6	7	8	9

——— Cut Line
——— Valley Fold

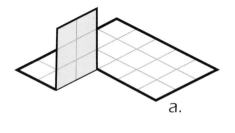

a.

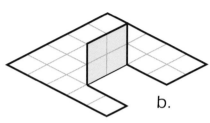

b.

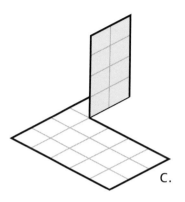

c.

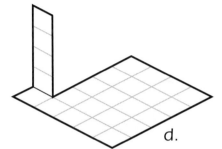

d.

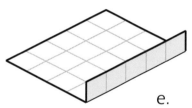

e.

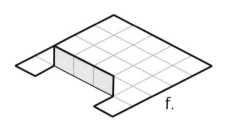

f.

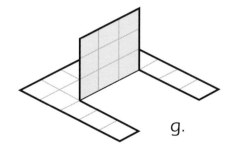

g.

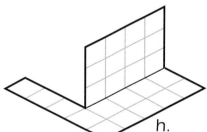

h.

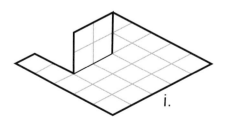

i.

A	a	b	c	d	e	f	g	h	i

Puzzle B
Plan views

In your imagination cut and fold these squares and match them to the views opposite.

1.

2.

3.

4.

5.

6.

7.

8.

9.

B	1	2	3	4	5	6	7	8	9

——— Cut Line
——— Valley Fold

Folded views

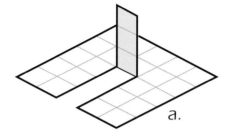

a.

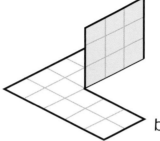

b.

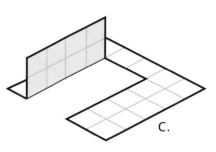

c.

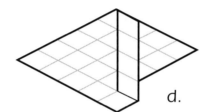

d.

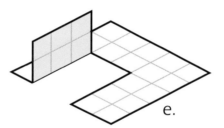

e.

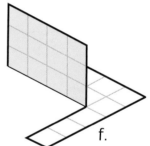

f.

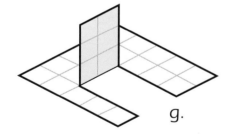

g.

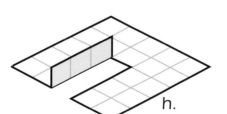

h.

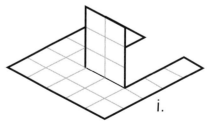

i.

B	a	b	c	d	e	f	g	h	i

Puzzle C
Plan views

In your imagination cut and fold these squares and match them to the views opposite.

1.

2.

3.

4.

5.

6.

7.

8.

9.

C	1	2	3	4	5	6	7	8	9

—— Cut Line

—— Valley Fold

Folded views

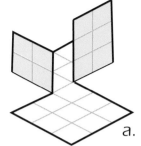

a.

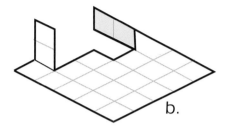

b.

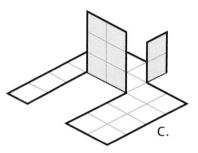

c.

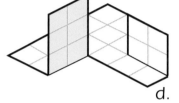

d.

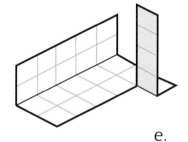

e.

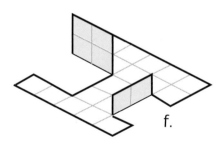

f.

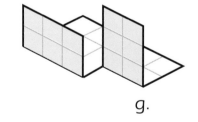

g.

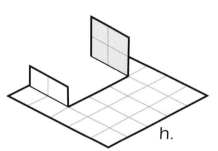

h.

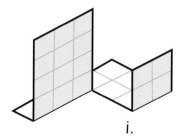

i.

C	a	b	c	d	e	f	g	h	i

Puzzle D
Plan views

In your imagination cut and fold these squares and match them to the views opposite.

1.

2.

3.

4.

5.

6.

7.

8.

9.

D	1	2	3	4	5	6	7	8	9

———— Cut Line
———— Valley Fold

Folded views

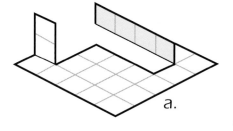

a.

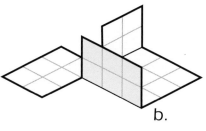

b.

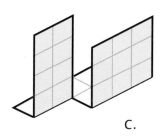

c.

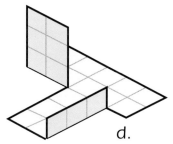

d.

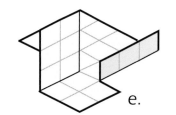

e.

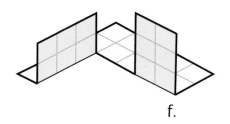

f.

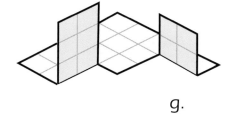

g.

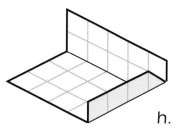

h.

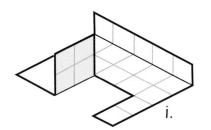

i.

D	a	b	c	d	e	f	g	h	i	

Puzzle E
Plan views

In your imagination cut and fold these squares and match them to the views opposite.

1.

2.

3.

4.

5.

6.

7.

8.

9.

E	1	2	3	4	5	6	7	8	9

———— Cut Line

———— Valley Fold

Folded views

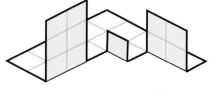

a.

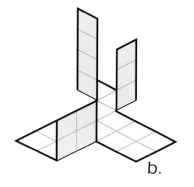

b.

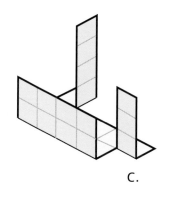

c.

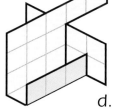

d.

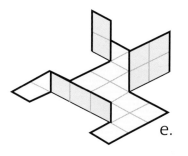

e.

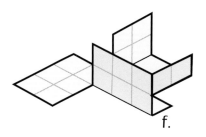

f.

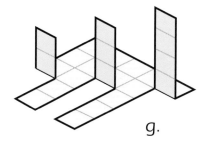

g.

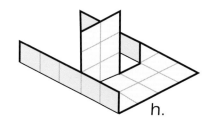

h.

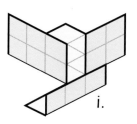

i.

E	a	b	c	d	e	f	g	h	i

Puzzle F
Plan views

In your imagination cut and fold these squares and match them to the views opposite.

1.

2.

3.

4.

5.

6.

7.

8.

9.

F	1	2	3	4	5	6	7	8	9

——— Cut Line
——— Valley Fold

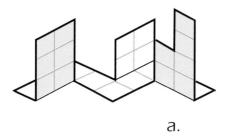

a.

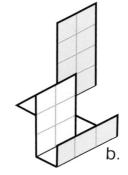

b.

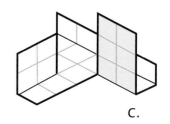

c.

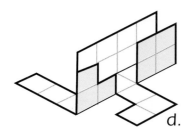

d.

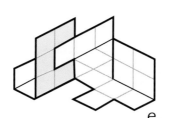

e.

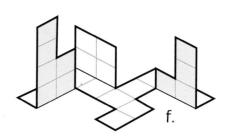

f.

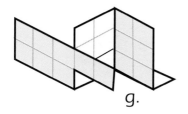

g.

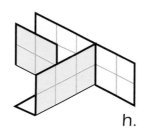

h.

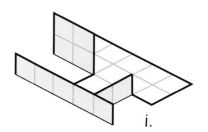

i.

F	a	b	c	d	e	f	g	h	i

Puzzle G
Plan views

In your imagination cut and fold these squares and match them to the views opposite.

1.

2.

3.

4.

5.

6.

7.

8.

9.

G	1	2	3	4	5	6	7	8	9

——— Cut Line
——— Valley Fold

Folded views

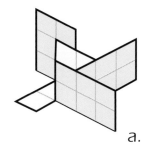

a.

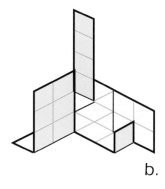

b.

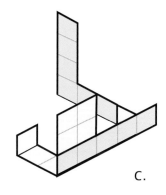

c.

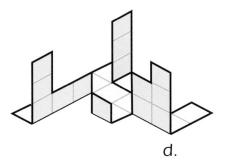

d.

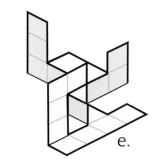

e.

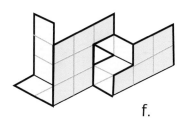

f.

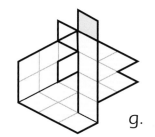

g.

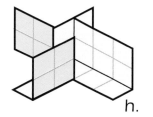

h.

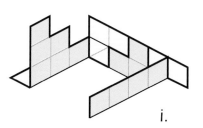

i.

G	a	b	c	d	e	f	g	h	i

Puzzle H
Plan views

In your imagination cut and fold these squares and match them to the views opposite.

1.

2.

3.

4.

5.

6.

7.

8.

9.

H	1	2	3	4	5	6	7	8	9

————— Cut Line
————— Valley Fold

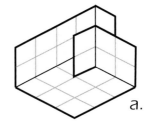

a.

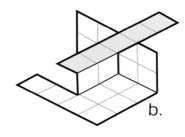

b.

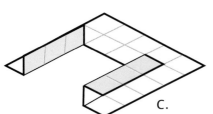

c.

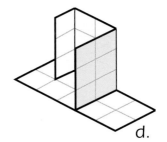

d.

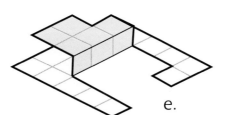

e.

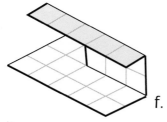

f.

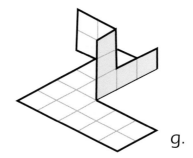

g.

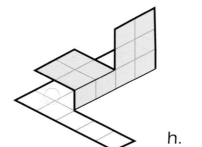

h.

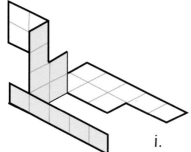

i.

H	a	b	c	d	e	f	g	h	i

Puzzle I
Plan views

In your imagination cut and fold these squares and match them to the views opposite.

1.

2.

3.

4.

5.

6.

7.

8.

9.

I	1	2	3	4	5	6	7	8	9

———— Cut Line
———— Valley Fold

Folded views

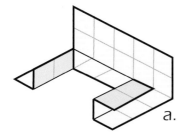

a.

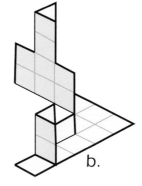

b.

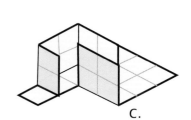

c.

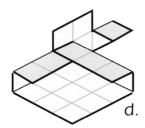

d.

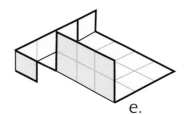

e.

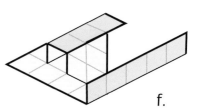

f.

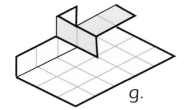

g.

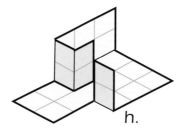

h.

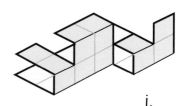

i.

I	a	b	c	d	e	f	g	h	i

Puzzle J
Plan views

In your imagination cut and fold these squares and match them to the views opposite.

1.

2.

3.

4.

5.

6.

7.

8.

9.

J	1	2	3	4	5	6	7	8	9

——— Cut Line
——— Valley Fold

Folded views

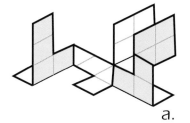

a.

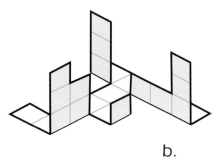

b.

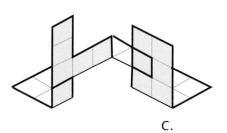

c.

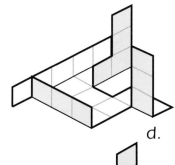

d.

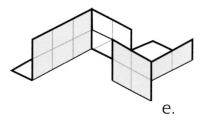

e.

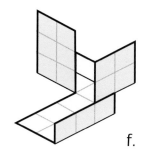

f.

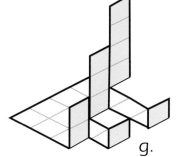

g.

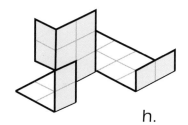

h.

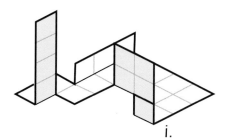

i.

J	a	b	c	d	e	f	g	h	i

Puzzle K
Plan views

In your imagination cut and fold these squares and match them to the views opposite.

1.

2.

3.

4.

5.

6.

7.

8.

9.

K	1	2	3	4	5	6	7	8	9

——— Cut Line
——— Valley Fold

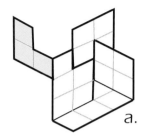

a.

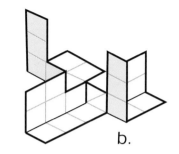

b.

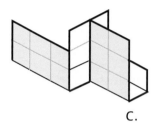

c.

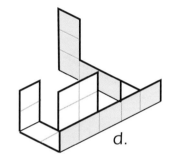

d.

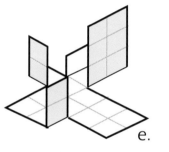

e.

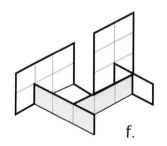

f.

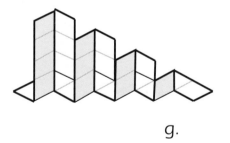

g.

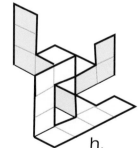

h.

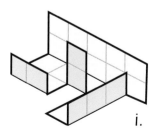

i.

K	a	b	c	d	e	f	g	h	i

Puzzle L
Plan views

In your imagination cut and fold these squares and match them to the views opposite.

1.

2.

3.

4.

5.

6.

7.

8.

9.

L	1	2	3	4	5	6	7	8	9

———— Cut Line
———— Valley Fold
———— Hill Fold

Folded views

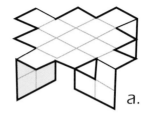

a.

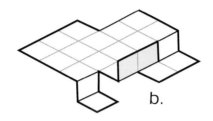

b.

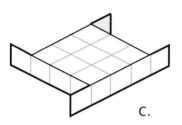

c.

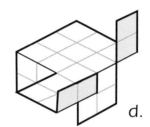

d.

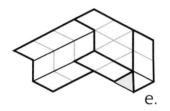

e.

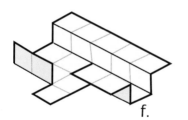

f.

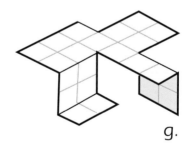

g.

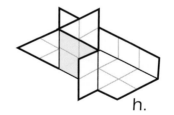

h.

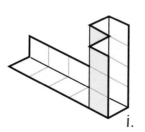

i.

L	a	b	c	d	e	f	g	h	i

Puzzle M
Plan views

In your imagination cut and fold these squares and match them to the views opposite.

1.

2.

3.

4.

5.

6.

7.

8.

9.

M	1	2	3	4	5	6	7	8	9

—— Cut Line
—— Valley Fold
—— Hill Fold

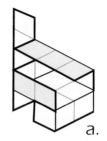

a.

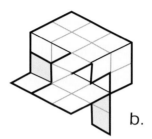

b.

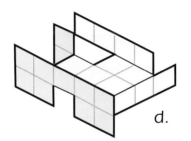

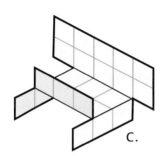

c.

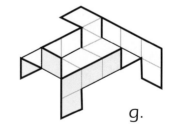

d.

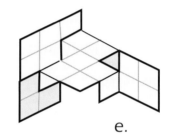

e.

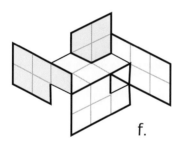

f.

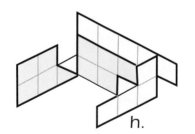

g.

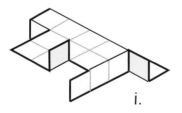

h.

i.

M	a	b	c	d	e	f	g	h	i

Master
Plan views

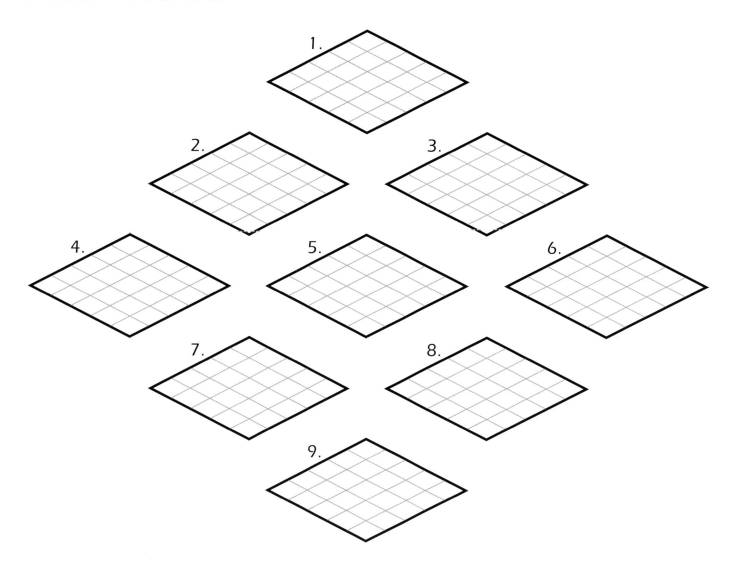

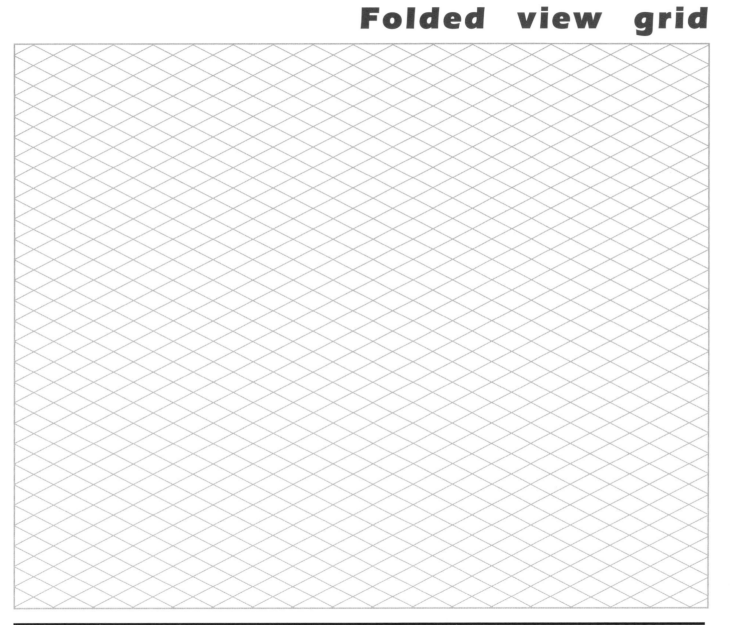

Master
Squares to cut

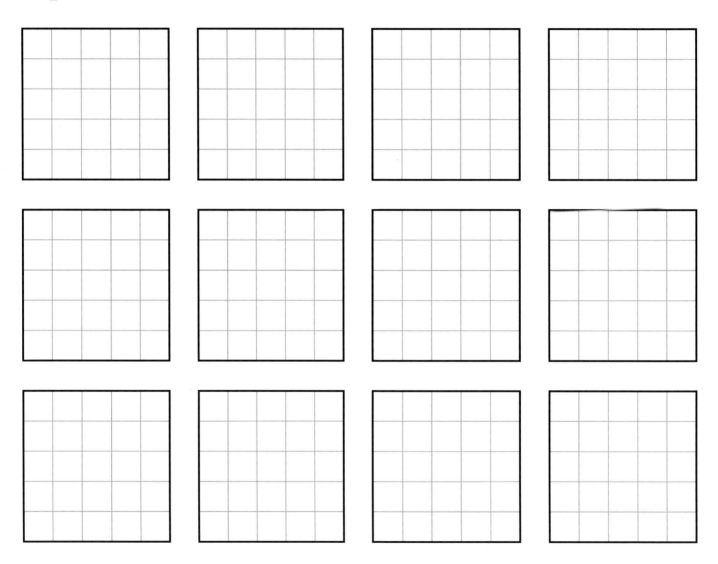